Bayron Ruiz

COMPOSITION, RICHNESS AND FOREST PRODUCTIVITY OF A TROPICAL FOREST

Bayron Ruiz

COMPOSITION, RICHNESS AND FOREST PRODUCTIVITY OF A TROPICAL FOREST

Importance Value Index and Standing timber volume

ScienciaScripts

Imprint
Any brand names and product names mentioned in this book are subject to trademark, brand or patent protection and are trademarks or registered trademarks of their respective holders. The use of brand names, product names, common names, trade names, product descriptions etc. even without a particular marking in this work is in no way to be construed to mean that such names may be regarded as unrestricted in respect of trademark and brand protection legislation and could thus be used by anyone.

Cover image: www.ingimage.com

This book is a translation from the original published under ISBN 978-613-9-46779-2.

Publisher:
Sciencia Scripts
is a trademark of
Dodo Books Indian Ocean Ltd. and OmniScriptum S.R.L publishing group

120 High Road, East Finchley, London, N2 9ED, United Kingdom
Str. Armeneasca 28/1, office 1, Chisinau MD-2012, Republic of Moldova, Europe
Managing Directors: Ieva Konstantinova, Victoria Ursu
info@omniscriptum.com

Printed at: see last page
ISBN: 978-620-8-51231-6

CONTENTS

SUMMARY

By means of a previous study of the municipality of the canton of San Pablo, the amount of forest mass that makes up a part of this municipality was determined by means of the Gentry method (1982), as this is a technique that determines with greater precision the amount of timber and non-timber species that make up this type of forest.

The vertical and horizontal structure of the analysed forest will be known, and the respective conclusions will be drawn for each index, as well as the species that for some reason predominate in the examined area.

1. INTRODUCTION

The forest ecosystems of the tropical region correspond to the most diverse biological complexes of the biosphere, whose services meet the needs of society and the human groups that live there in terms of timber, firewood, fibre, medicine, wildlife that provides animal protein, climate and water regulation, which makes it an invaluable system for man, as a basis for the sustainability of life.

In accordance with the above, any strategy generated for their sustainable management must be based on knowledge of both their form and their functioning, which guarantees the shaping of biodiversity and the rational use of their services. It is therefore essential that the officials in charge of their administration know and handle the functional elements for the technical and scientific evaluation of these forest masses, which will enable decisions to be taken for their management, conservation and recovery.

This document is a valuable tool that the university's teachers have wanted to contribute to broaden the knowledge of the country's forest ecosystems, which is presented as a result of the research carried out over the last few decades in the different forest ecosystems of the department and the whole country.

2. OBJECTIVES

GENERAL

- To analyse and determine the existing floristic composition in a given area of the municipality of Cantón del San Pablo-Chocó.

SPECIFIC

- Identify existing timber and non-timber species in the municipality of Canton San Pablo.

- Construct and analyse the forest profile diagram, based on the information obtained in the working area, taking into account the site with the highest abundance of species.

- Assess and analyse the vertical and horizontal structure of the inventoried forest.

3. METHODOLOGY

3.1.1 STUDY AREA

The municipality of Cantón de San Pablo is located in the central southern part of the department of Choco, about 70 km by road from Quibdó, the capital of the department, more specifically on the isthmus of San Pablo. Its geography is crossed by the Pan-American highway that leads to the port of Tribuga on the Colombian Pacific Ocean.

3.1.2 GEOGRAPHICAL POSITION

Located on the right bank of the San Pablo River at 05° 20' 20" north longitude and 76° 43' 53" west longitude.

3.1.3 EXTENT AND LIMITS

It has an extension of 386 Km^2 which represents 0.8% of the departmental surface. It is bordered to the north by the municipality of Río Quito and Certegui, to the south by the municipality of Istmina, to the east by the municipality of Unión Panamericana and to the west by the municipalities of Alto Baudó and Medio Baudó.

3.1.4 LOCATION OF POPULATION

It has a population of 10,890 inhabitants approx. distributed in its 7 townships: Managrú, Puerto pervel, Tarido, La Victoria, Pavaza, Boca de raspadura and Guapango its veredas are: Boca de Jorodó, San José de quité, la Isla, Zutana, Truadó and Puerto Juan.

3.1.5 ECONOMY

The fertility of the soils allows it to project itself as the second agricultural larder of the department, after Baudó. The breeding of large and small species is a prominent source of income for the communities, as is the commercialisation of timber; mining is carried out on a very small scale. The most harvested products are: Plantain, rice, yucca, and avocado, borojó, lemon and chontaduro fruit trees.

3.1.6 NATURAL CHARACTERISTICS

Due to its flat and jungle characteristics, it is very attractive for eco-tourism, with areas such as Taridó, La Victoria, Chagarapá and Puerto Pervel, among others, standing out.

3.1.7 ACCESS ROUTES

he main access road to the municipality is the Pan-American Highway, built by two canals that lead to Managrú, the municipal capital, and to Puerto Pervel. There are approximately 12 km of unpaved road, filled with stones, to both destinations, on which speeds of 40-60 km/h can be reached. Another important route is the Fluvial one that constitutes the rivers: Tiradó, San Pablo, Managrú, inside these ways the products of the most productive zones are taken out.

3.1.8 TOPOGRAPHY

The territory is strongly undulating, with dissections in some regions, and its altitude does not exceed 57 metres above sea level.

3.1.9 CLIMATOLOGY

It has an average annual temperature of 28ºc., regulated by the inter-oceanic winds and the fluvial currents of the rivers that bathe the municipality. Its lands are distributed in the warm thermal floor with an average humidity of 80% - 90%, loaded with humid air, which comes from the surrounding forest areas, which affects a cloudiness that varies from 4 - 6 hours / light, and an average annual rainfall of 6,000 mm / year, leading to a high diversity corresponding to the life zone of the tropical rainforest (bp -T), according to Holdridgee's classification.

3.1.10 VEGETATION There are variations in the composition, density and distribution of the area, with representative species such as: Chana, Pala Perico, Nuanamo, Caucho and Algarrobo. Among the species of economic importance.

3.1.11 HYDROGRAPHY

It has a great hydrographic wealth, which allows its inhabitants to move around easily. Among the most important rivers are: San Pablo, Arteria principal, Taridó, Managrúcito, Raspadura, Tuado, Managrú adentro and Chagarapá.

3.1.12 Soils

The municipality has poorly drained soils with low exchangeable base saturation (Gleysole districos - GLD), and soils originated from recent alluvial materials, with high exchangeable base saturation (Eutric Fluvisols - FLE), associated with poorly drained soils and high saturation of exchangeable bases (Eutric Fluvisols - FLE), and soils with argic horizon, low saturation of exchangeable bases and high iron oxide content (Ferric Acrisols - ACF), medium textures, gentle slopes located in plains and valleys.

3.2 ANALYSIS OF THE EXISTING FLORISTIC COMPOSITION IN THE MUNICIPALITY OF THE CANTON OF SAN PABLO-CHOCO

Through the intervention in the field, in the municipality of Cantó del San Pablo, the existing species in the area were previously quantified and grouped according to the number of the predominant plot.

4. SAMPLING DESIGN

4.1.1 Design

The methodology of Gentry (1982) was effectively used, which postulates that plots of 2 x 50m (total: 10) should be designed in which all species with a DBH > 1 cm should be measured. In addition, he suggests that, for the determination of the horizontal structure of a forest, the most representative site of the area should be selected. For the measurement of the plots and respective quantifications, the following criteria were taken into account:

- DAP.
- Total height of each tree.
- Commercial height
- Remarks (tree defects)
- Supporting table.
- Field form.
- Diameter tape or Forcípula
- Measuring tape 20 m or 50 m.
- Stakes.
- Machete.
- Pencils.
- Paper.

Subsequently, the vertical profile of the forest was drawn by representing the trees in the plot, considering the architecture of the crowns and the shape of the trunks,

The horizontal and vertical profile diagrams of the forest were drawn in pencil on chart paper during the survey of the area.

Basal Area

To calculate the basal area of the plot, the basal area of the stems of the sampled trees is summed; it is calculated for each tree as follows:

$$AB = nDAP^2/4$$

Where:

DBH = Diameter of stem measured at breast height.

AB = Basal area (m2).

= 31416,n

IMPORTANCE VALUE INDEX, MIXING RATIO AND AGGREGATE VALUE OF TREE SPECIES IN A FOREST.

The Importance Value Index (I.I.V.I.) is a value that numerically expresses the importance of a given species within a forest community. The I.V.I. of a species is calculated through abundance, frequency and dominance expressed in percentages, the Family Importance Value Index.

DEGREE OF AGGREGATION OF SPECIES:

It determines the spatial distribution of species. The interpretation of the degree of aggregation of the sum of their relative values of the hundred is made taking into account the following parameters:

- **Ga > a 1**. indicates a tendency to cluster.
- **Ga > a 2**, indicates that the species has a clumped distribution
- **Ga < a 1**, indicates that the species is dispersed.

ANALYSIS OF THE RESULTS

Mixing ratio

C= number of species/ number of individuals

C= 35/218=1/6

This means that for every one species there are 6 individuals

Density

D= total number of individuals/total area of the sampling area

D=218/1000m²= 0.218

Floristic composition of the sampling carried out in the tropical rainforest of Managrú:

It was noted that there was a great variety of species and families, the most representative being the following: Arecaceae 63, Apocynaceae 17, Melastomataceae 17, Burseraceae 15, Moraceae 12 and Sapotaceae 11 individuals **(see table 2).**

Calculation of relative and absolute abundance.

The highest number of trees per species recorded in each sampling unit corresponds to the meme palm with 33 individuals and a percentage of 15.138%; followed by the lily with 17 individuals and a percentage of 7.798%, the ant with 17 individuals per species and a percentage of 7.798%. Anime with 15 individuals and a percentage of 6.8881% and carob with 12 individuals and a percentage of 5.505% **(see table 3).**

Calculation of absolute and relative frequency

The species with the highest incidence in each of the sampling units were: class I with 155 individuals and class II with 43 individuals, with the following species being the most predominant in each class: meme palm 8.264%, lily 7.438%, ant 5.785%, anime 5.785% and carob 4.955%. This leads to the conclusion that the forest is very heterogeneous **(see table 4).**

Calculation of absolute and relative dominance.

The species with the highest degree of cover as an expression of the space occupied by them; and with a sum of the basic areas of the same species present within each of the sampling units were: sawdust 22.128%, iris 11.589%, choiba 11.327%, anime 9.064% and carob 6.089% **(see table 5).**

Calculation of the degree of aggregation

It determines the spatial distribution of the species, which indicates the tendency for clustering, clustered distribution and dispensation. The species most prone to clustering are:

2,18543453	caimito colorado	2,06377023	palma mil peso
2,18543453	charcoal burner	1,93670887	Guasca hoji wide
2,18543453	carra	1,93670887	Guasco colorado
2,18543453	helmet	1,93670887	bitter palm
2,18543453	coroba	1,80303022	Churimo
2,18543453	red guasimo	1,80303022	Soap
2,18543453	Jagua	1,80303022	taparo palm
2,18543453	white wool	1,80303022	swamp
2,18543453	chacarra palm	1,66096405	knife
2,18543453	palo blanco	1,66096405	milkman
2,06377023	milkweed	1,66096405	pangana palm
2,06377023	cargadero colorado	1,66096405	zancona palm
2,06377023	Choiba	1,50776496	Carob
2,06377023	black jigua	1,50776496	Caimito white
2,06377023	pot-bellied palm	1,3387425	Anime
2,06377023	palma don pedrito	1,3387425	Sawdust

(See table 7).

Calculation of structural analysis of sampling

Quantitative indicators such as the number of trees per species, density, abundance, frequency, dominance and the importance value index were used. The latter had a maximum value of 300, the most representative being: sawdust 32,500, meme palm 28,925, lily 26,825, anime 21,730 and carob tree 16,553 **(see table 6).**

Existence of species by diameter class

The highest number of individuals (155) was found in the Class I range. In class II (43). Class III (12), Class IV (5), Class V (2) and VI (1) (**see table 8**).

Diametric distributions

Through the analysis of this forest, it was possible to deduce that it has been a forest that has been heavily intervened, in which most of the species are in a state of natural regeneration, in class I (155 individuals), and the total volume of species found in the forest is 24.0816514 m^3, which is very low for a total use of this forest. **(See table 13).**

Analysis of the ogawa diagram

According to the distribution of the dots in the graph, it means that the species are conglomerated, forming a very homogeneous swarm of bees (see **Graphical analysis of the ogawa diagram).**

5. CONCLUSION

Through the execution of this practice it was possible to deduce that:

- It is important to know the diversity of our forests, as this is often a very feasible alternative to obtain income from previous harvesting.

- In a few years' time, the forest under analysis will be a very important source of income in terms of exploitation, as the species are generally in a state of natural regeneration.

Recommendations

Some species should be harvested as quickly as possible, as they are rotting due to their advanced stage of development, and the forest should be released so that the most predominant species can develop well.

6. BIBLIOGRAPHY

- FOREST MANAGEMENT BASED ON NATURAL FOREST REGENERATION, a case study in the high-altitude oak forests of the Salamanca mountain range, Costa Rica.

- FUNDAMENTALS AND METHODOLOGY FOR PLANT IDENTIFICATION by Gilberto Emilio Maecha vega, BIOPACIFICO project, ministry of environment UNDP - GEF February 1997 Santa Fe de Bogotá D.C- COLOMBIA

ANNEXES

Table: Floristic composition of the sampling conducted in the tropical rainforest 2.

NO.	Family	Scientific name	Common Name	No. trees
1	ANONACEAE	*Anaxagorea clavata*	coroba	1
2	ANONACEAE	*Guatteria cuatrecasasii*	cargadero colorado	2
3	APOCYNACEAE	*Couma macrocarpa*	lily	17
4	ARALIACEAE	*Dendropanax sp*	palo blanco	1
5	ARECACEAE	*Welfia georgii*	bitter palm	5
6	ARECACEAE	*Triartea delfoidea*	pot-bellied palm	3
7	ARECACEAE	*Oenocarpus mapora*	palma don pedrito	2
8	ARECACEAE	*Wettinia quinaria*	palm meme	33
9	ARECACEAE	*Oenocarpus bataua*	palma mil peso	2
10	ARECACEAE	*Rhaphia taedigera*	pangana palm	7
11	ARECACEAE	*Attalea amiggdalina*	taparo palm	4
12	ARECACEAE	*Socrotea exorrhiza*	zancona palm	6
13	ARECACEAE	*Bactris chascarra*	chacarra palm	1
14	BOMBACACACEAE	*Humberodendron patinoi*	carra	1
15	BOMBACACACEAE	*Bombax sp*	white wool	2
16	BURCERACEAE	*Protium veneralense*	Anime	15
17	CAESALPINACEAE	*Himenaea oblongifolia*	Carob	12
18	CHRYSOBALANACEAE	*Licania durifolia*	charcoal burner	1
19	EUPHORBIACEA	*Hieronyma chocoensis*	swamp	5
20	EUPHORBIACEAE	*Croton lillipianus*	milkweed	2
21	FABACEAE	*Dipteris panamensis*	Choiba	6
22	LAURACEAE	*Ocotea cernua*	black jigua	3
23	LECYTHIDACEAE	*Eshweilera sclerophylla*	Guasca hoji wide	9
24	LECYTHIDACEAE	*Eshweilera oligosperma*	Guasco colorado	4
25	MELASTOMATACEAE	*Miconia sp*	Ant	17
26	MIMOSACEAE	*Parkia oppositifolia*	Sawdust	10
27	MIMOSACEAE	*Inga edulis*	Churimo	7
28	MORACEA	*Clarisia biflora*	knife	6

29	MORACEA	*Brosimun utile*	milkman	12
30	OCHNACEAE	*Cespedesia macrophyla*	helmet	1
31	RUBIACEAE	*Genipa americana*	Jagua	1
32	SAPINDACEAE	*Isertia pittieri*	Soap	5
33	SAPOTACEAE	*Chrysophyllum auratum*	Caimito white	11
34	SAPOTACEAE	*Chrysophyllum sp*	caimito colorado	2
35	TILIACEAE	*Luehea seemannii*	red guasimo	2
	Total			**218**

Table: 3 Calculation of Abundance. Absolute and Relative Abundance of Sampling in Tropical Rainforest

NO.	Species	Number of trees per sampling unit											Aa	Ar(%)
		1	2	3	4	5	6	7	8	9	10	Total		
1	Carob	2	3	1	2	1					3	12	12	5,505
2	milkweed								1	1		2	2	0,917
3	Anime	5				4	2	1	1	1	1	15	15	6,881
4	Sawdust	2			1	1		1	1	1	3	10	10	4,587
5	Caimito white	2	2	1	3		2	1				11	11	5,046
6	caimito colorado	2										2	2	0,917
7	charcoal burner								1			1	1	0,459
8	cargadero colorado				1					1		2	2	0,917
9	carra								1			1	1	0,459
10	helmet										1	1	1	0,459
11	knife			2			1	1		1	1	6	6	2,752
12	Choiba	5		1								6	6	2,752
13	Churimo				1	3		1		2		7	7	3,211
14	coroba	1										1	1	0,459
15	Guasca hoji wide				3	3			3			9	9	4,128
16	Guasco colorado	1			2		1					4	4	1,835
17	red guasimo									2		2	2	0,917
18	Ant	3	3	4	3		1	2		1		17	17	7,798
19	Soap				1	1			1		2	5	5	2,294
20	Jagua		1									1	1	0,459
21	black jigua			1					2			3	3	1,376
22	white wool								2			2	2	0,917
23	milkman		4	2		2	2	2				12	12	5,505
24	lily	2	2	3	1	1		2	3	2	1	17	17	7,798
25	bitter palm						1	3			1	5	5	2,294

26	pot-bellied palm					2				1		3	3	1,376
27	chacarra palm		1									1	1	0,459
28	palma don pedrito						1				1	2	2	0,917
29	palm meme	3	3	3	4	5	2	2	1	4	6	33	33	15,138
30	palma mil peso			1					1			2	2	0,917
31	pangana palm	1	1		1		2	2				7	7	3,211
32	taparo palm		1	1		1	1					4	4	1,835
33	zancona palm			1		2	1		1	1		6	6	2,752
34	palo blanco						1					1	1	0,459
35	swamp	1						1	2	1		5	5	2,294
	Total	**30**	**21**	**21**	**23**	**26**	**18**	**19**	**21**	**19**	**20**	**218**	**218**	**100**

Table: 4 Calculation of Frequency. Absolute and relative frequency of sampling in the tropical rainforest.

NO.	Species	No. of sampling units where the species occurs													Frequency classes						
		1	2	3	4	5	6	7	8	9	10	Total	Fa	Fr(%)	I	II	III	IV	V	VI	Total
1	Carob	x	x	x	x	x					x	6	60	4,959	7	3	2				12
2	milkweed								x	x		2	20	1,653	1		1				2
3	Anime	x				x	x	x	x	x	x	7	70	5,785	9	3	3				15
4	Sawdust	x			x	x		x	x	x	x	7	70	5,785	1	4	1	2	2		10
5	Caimito white	x	x	x	x		x	x				6	60	4,959	10	1					11
6	caimito colorado	x										1	10	0,826	2						2
7	charcoal burner								x			1	10	0,826	1						1
8	cargadero colorado				x					x		2	20	1,653	2						2
9	carra								x			1	10	0,826		1					1
10	helmet										x	1	10	0,826	1						1
11	knife			x			x	x		x	x	5	50	4,132	5	1					6
12	Choiba	x		x								2	20	1,653	3	2				1	6
13	Churimo				x	x		x		x		4	40	3,306	5		1	1			7
14	coroba	x										1	10	0,826	1						1
15	Guasca hoji wide				x	x			x			3	30	2,479	8	1					9
16	Guasco colorado	x			x		x					3	30	2,479	3	1					4
17	red guasimo									x		1	10	0,826	1	1					2
18	Ant	x	x	x	x		x	x		x		7	70	5,785	16	1					17
19	Soap				x	x			x		x	4	40	3,306	2	2		1			5
20	Jagua		x									1	10	0,826	1						1
21	black jigua			x					x			2	20	1,653	3						3
22	white wool								x			1	10	0,826	2						2

23	milkman		x	x		x	x	x				5	50	4,132	7	4	1				12
24	lily	x	x	x	x	x		x	x	x	x	9	90	7,438	9	5	2	1			17
25	bitter palm						x	x			x	3	30	2,479	4	1					5
26	pot-bellied palm					x				x		2	20	1,653	2	1					3
27	chacarra palm		x									1	10	0,826	1						1
28	palma don pedrito						x				x	2	20	1,653	1	1					2
29	palm meme	x	x	x	x	x	x	x	x	x	x	10	100	8,264	28	5					33
30	palma mil peso			x					x			2	20	1,653		1	1				2
31	pangana palm	x	x		x		x	x				5	50	4,132	6	1					7
32	taparo palm		x	x		x	x					4	40	3,306	4						4
33	zancona palm			x		x	x		x	x		5	50	4,132	4	2					6
34	palo blanco						x					1	10	0,826	1						1
35	swamp	x						x	x	x		4	40	3,306	4	1					5
	Total	**13**	**10**	**12**	**12**	**12**	**13**	**12**	**14**	**13**	**10**	**121**	**1210**	**100**	**155**	**43**	**12**	**5**	**2**	**1**	218

Table: 5 Calculation of Dominance. Absolute and relative of the sampling carried out.

in the tropical rainforest

NO.	Species	Basal area in m2 per sampling unit												
		1	2	3	4	5	6	7	8	9	10	total	Da	Dr(%)
1	Carob	0,0228	0,0134	0,045	0,0086	3E-04					0,054	0,144	0,144	6,089
2	milkweed								0,0003	0,031		0,032	0,032	1,341
3	Anime	0,0426				0,128	0,015	0,006	0,0177	0,005	8E-05	0,215	0,215	9,064
4	Sawdust	0,012			0,0227	0,071		0,139	0,0314	0,126	0,123	0,524	0,524	22,128
5	Caimito white	0,0032	0,0058	7E-04	0,007		0,022	8E-05				0,039	0,039	1,660
6	caimito colorado	0,001										0,001	0,001	0,043
7	charcoal burner								0,0013			0,001	0,001	0,053
8	cargadero colorado				8E-05					0,004		0,004	0,004	0,166
9	carra								0,0079			0,008	0,008	0,332
10	helmet										8E-05	0,000	0,000	0,003
11	knife			0,013			8E-05	0,002		3E-04	1E-03	0,017	0,017	0,701
12	Choiba	0,2586		0,01								0,268	0,268	11,327
13	Churimo				0,005	0,003		0,091		0,035		0,134	0,134	5,648
14	coroba	0,0003										0,000	0,000	0,013
15	Guasca hoji wide				0,0078	0,015			0,007			0,029	0,029	1,241
16	Guasco colorado	0,0003			0,010		7E-04					0,011	0,011	0,458
17	red guasimo									0,029		0,029	0,029	1,228

18	Ant	0,0108	0,0108	0,007	0,0015		3E-04	0,001		3E-04		0,032	0,032	1,355
19	Soap				0,0908	0,020			0,000		0,014	0,125	0,125	5,281
20	Jagua		0,0013									0,001	0,001	0,053
21	black jigua			0					0,003			0,003	0,003	0,126
22	white wool								0,0092			0,009	0,009	0,388
23	milkman		0,051	0,026		0,032	0,007	0,005				0,121	0,121	5,117
24	lily	0,0283	0,0086	0,042	0,108	0,003		0,027	0,0324	0,025	1E-03	0,274	0,274	11,589
25	bitter palm						3E-04	0,003			0,018	0,021	0,021	0,903
26	pot-bellied palm					0,01				0,003		0,012	0,012	0,524
27	chacarra palm		0,002									0,002	0,002	0,083
28	palma don pedrito						0,006				0,008	0,014	0,014	0,601
29	palm meme	0,0092	0,012	0,009	0,0189	0,04	0,006	0,001	0,013	0,008	0,013	0,131	0,131	5,523
30	palma mil peso			0,066					0,0201			0,086	0,086	3,640
31	pangana palm	0,0007	0,018		0,000		0,005	4E-04				0,024	0,024	1,030
32	taparo palm		0,002	7E-04		0,001	0,000					0,003	0,003	0,146
33	zancona palm			7E-04		0,008	0,011		8E-05	0,003		0,023	0,023	0,967
34	palo blanco						0,006					0,006	0,006	0,269
35	swamp	0,0113						0,006	0,0031	7E-04		0,022	0,022	0,909
	Total	**0,4009**	**0,1244**	**0,2205**	**0,2804**	**0,3301**	**0,0804**	**0,2823**	**0,1468**	**0,2696**	**0,2312**	**2,3668**	**2,3668**	**100**

Table: 6 Importance Value Index (IVI)

Species		No. of	Density	Abundance		Frequency		Dominance		I V I
Order No.	Name	Trees	(D)	Aa	Ar(%)	Fa	Fr(%)	Da	Dr(%)	
1	Carob	12	1,2	12	5,505	60	4,959	0,144	6,0893	16,553
2	milkweed	2	0,2	2	0,9174	20	1,65289	0,0317	1,3406	3,911
3	Anime	15	1,5	15	6,8807	70	5,78512	0,2145	9,0642	21,730
4	Sawdust	10	1	10	4,5872	70	5,78512	0,5237	22,128	32,500
5	Caimito white	11	1,1	11	5,0459	60	4,95868	0,0393	1,6602	11,665
6	caimito colorado	2	0,2	2	0,9174	10	0,82645	0,001	0,0431	1,787
7	charcoal burner	1	0,1	1	0,4587	10	0,82645	0,0013	0,0531	1,338
8	cargadero colorado	2	0,2	2	0,9174	20	1,65289	0,0039	0,1659	2,736
9	carra	1	0,1	1	0,4587	10	0,82645	0,0079	0,3318	1,617
10	helmet	1	0,1	1	0,4587	10	0,82645	8E-05	0,0033	1,288
11	knife	6	0,6	6	2,7523	50	4,13223	0,0166	0,701	7,586
12	Choiba	6	0,6	6	2,7523	20	1,65289	0,2681	11,327	15,733
13	Churimo	7	0,7	7	3,211	40	3,30579	0,1337	5,6479	12,165
14	coroba	1	0,1	1	0,4587	10	0,82645	0,0003	0,0133	1,298
15	Guasca hoji wide	9	0,9	9	4,1284	30	2,47934	0,0294	1,2411	7,849
16	Guasco colorado	4	0,4	4	1,8349	30	2,47934	0,0108	0,4579	4,772
17	red guasimo	2	0,2	2	0,9174	10	0,82645	0,0291	1,2278	2,972
18	Ant	17	1,7	17	7,7982	70	5,78512	0,0321	1,3547	14,938
19	Soap	5	0,5	5	2,2936	40	3,30579	0,125	5,281	10,880
20	Jagua	1	0,1	1	0,4587	10	0,82645	0,0013	0,0531	1,338
21	black jigua	3	0,3	3	1,3761	20	1,65289	0,003	0,1261	3,155

22	white wool	2	0,2	2	0,9174	10	0,82645	0,0092	0,3883	2,132
23	milkman	12	1,2	12	5,5046	50	4,13223	0,1211	5,117	14,754
24	lily	17	1,7	17	7,7982	90	7,43802	0,2743	11,589	26,825
25	bitter palm	5	0,5	5	2,2936	30	2,47934	0,0214	0,9026	5,676
26	pot-bellied palm	3	0,3	3	1,3761	20	1,65289	0,0124	0,5243	3,553
27	chacarra palm	1	0,1	1	0,4587	10	0,82645	0,002	0,083	1,368
28	palma don pedrito	2	0,2	2	0,9174	20	1,65289	0,0142	0,6006	3,171
29	palm meme	33	3,3	33	15,138	100	8,26446	0,1307	5,5227	28,925
30	palma mil peso	2	0,2	2	0,9174	20	1,65289	0,0862	3,6403	6,211
31	pangana palm	7	0,7	7	3,211	50	4,13223	0,0244	1,0304	8,374
32	taparo palm	4	0,4	4	1,8349	40	3,30579	0,0035	0,146	5,287
33	zancona palm	6	0,6	6	2,7523	50	4,13223	0,0229	0,9671	7,852
34	palo blanco	1	0,1	1	0,4587	10	0,82645	0,0064	0,2688	1,554
35	swamp	5	0,5	5	2,2936	40	3,30579	0,0215	0,9092	6,509
	Total	**218**	**21,8**	**218**	**100**	**1210**	**100**	**2,367**	**100**	**300**

Table 7. Calculation of the Degree of Aggregation for Tropical Rainforest Sampling

NO.	Species	No. of trees of sampling in which the species occurs	Number of Trees by Species	Frequency Absolute	Density Expected (De)	Density Observed (Ob)	Degree of Aggregation G.A.
1	Carob	6	12	60	0,398	0,6	1,508
2	milkweed	2	2	20	0,097	0,2	2,064
3	Anime	7	15	70	0,523	0,7	1,339
4	Sawdust	7	10	70	0,523	0,7	1,339
5	Caimito white	6	11	60	0,398	0,6	1,508
6	caimito colorado	1	2	10	0,046	0,1	2,185
7	charcoal burner	1	1	10	0,046	0,1	2,185
8	cargadero colorado	2	2	20	0,097	0,2	2,064
9	carra	1	1	10	0,046	0,1	2,185
10	helmet	1	1	10	0,046	0,1	2,185
11	knife	5	6	50	0,301	0,5	1,661
12	Choiba	2	6	20	0,097	0,2	2,064
13	Churimo	4	7	40	0,222	0,4	1,803
14	coroba	1	1	10	0,046	0,1	2,185
15	Guasca hoji wide	3	9	30	0,155	0,3	1,937
16	Guasco colorado	3	4	30	0,155	0,3	1,937
17	red guasimo	1	2	10	0,046	0,1	2,185
18	Ant	7	17	70	0,523	0,7	1,339
19	Soap	4	5	40	0,222	0,4	1,803
20	Jagua	1	1	10	0,046	0,1	2,185

21	black jigua		2	3	20	0,097	0,2	2,064
22	white wool		1	2	10	0,046	0,1	2,185
23	milkman		5	12	50	0,301	0,5	1,661
24	lily		9	17	90	1,000	0,9	0,900
25	bitter palm		3	5	30	0,155	0,3	1,937
26	pot-bellied palm		2	3	20	0,097	0,2	2,064
27	chacarra palm		1	1	10	0,046	0,1	2,185
28	palma don pedrito		2	2	20	0,097	0,2	2,064
29	palm meme		10	33	100	3,000	1	0,333
30	palma mil peso		2	2	20	0,097	0,2	2,064
31	pangana palm		5	7	50	0,301	0,5	1,661
32	taparo palm		4	4	40	0,222	0,4	1,803
33	zancona palm		5	6	50	0,301	0,5	1,661
34	palo blanco		1	1	10	0,046	0,1	2,185
35	swamp		4	5	40	0,222	0,4	1,803
	Total		**121**	**218**	**1210**	**10,057**	**12,100**	**64,232**

Table 9. Timber volume by species

	Species	N° individual	Volume m3
1	Carob	12	1,321012012
2	milkweed	2	0,24553489
3	Anime	15	1,795501448
4	Sawdust	10	7,038819988
5	Caimito white	11	0,155906568
6	caimito colorado	2	0,002695493
7	charcoal burner	1	0,002391637
8	cargadero colorado	2	0,012080709
9	carra	1	0,02450448
10	helmet	1	3,67567E-05
11	knife	6	0,132881669
12	Choiba	6	3,491355428
13	Churimo	7	1,423955333
14	coroba	1	0,000294054
15	Guasca hoji wide	9	0,108677369
16	Guasco colorado	4	0,019995656
17	red guasimo	2	0,368302334
18	Ant	17	0,110512142
19	Soap	5	1,485260641
20	Jagua	1	0,003920717
21	black jigua	3	0,002793511
22	white wool	2	0,023156734
23	milkman	12	0,83364241

24	lily	17	2,537392958
25	bitter palm	5	0,22491437
26	pot-bellied palm	3	0,117082405
27	chacarra palm	1	0,002756754
28	palma don pedrito	2	0,170428658
29	palm meme	33	1,009553945
30	palma mil peso	2	0,775076702
31	pangana palm	7	0,118050332
32	taparo palm	4	0,005035671
33	zancona palm	6	0,278253271
34	palo blanco	1	0,069470201
35	swamp	5	0,170404154
	Total	**218**	**24,0816514**

Table 10. Volume per family

NO.	Family	#	Volume
1	ANONACEAE	2	0,01237476
2	APOCYNACEAE	1	2,53739296
3	ARALIACEAE	1	0,0694702
4	ARECACEAE	9	*2,70115211*
5	BOMBACACACEAE	2	0,04766121
6	BURCERACEAE	1	1,79550145
7	CAESALPINACEAE	1	1,32101201
8	CHRYSOBALANACEAE	1	0,00239164
9	EUPHORBIACEAE	2	0,41593904
10	FABACEAE	1	3,49135543
11	LAURACEAE	1	0,00279351
12	LECYTHIDACEAE	2	0,12867302
13	MELASTOMATACEAE	1	0,11051214
14	MIMOSACEAE	2	8,46277532
15	MORACEA	2	0,96652408
16	OCHNACEAE	1	3,6757E-05
17	RUBIACEAE	1	0,00392072
18	SAPINDACEAE	1	1,48526064

19	SAPOTACEAE	2	0,15860206
20	TILIACEAE	1	0,36830233
	TOTAL	**35**	**24,0816514**

Table 11. stratification of the analysed forest

Tree layer	Symbol	Height limit (m)	
Upper stratum (dominant)	It is	> 20	15
Middle stratum (Co-dominant)	Em	15-20	39
Lower stratum (Dominated)	Ei	< 15	164
TOTAL			**218**

Table 12. importance value index per family

NO. Order	FAMILY	No. of Species per family	Basal Area per family	IVIF
1	ANONACEAE	2	0,00424116	7,2696266
2	APOCYNACEAE	1	0,27428132	22,244025
3	ARALIACEAE	1	0,00636174	3,5846497
4	ARECACEAE	9	0,31755214	67,571614
5	BOMBACACACEAE	2	0,01704318	1,4404906
6	BURCERACEAE	1	0,21453201	18,802113
7	CAESALPINACEAE	1	0,1441209	14,451012
8	CHRYSOBALANACEAE	1	0,00125664	3,368953
9	EUPHORBIACEAE	2	0,05325012	11,175177
10	FABACEAE	1	0,26809629	16,936828
11	LAURACEAE	1	0,00298452	1,4059792
12	LECYTHIDACEAE	2	0,04021248	13,376614
13	MELASTOMATACEAE	1	0,03206396	12,010049
14	MIMOSACEAE	2	0,65739944	41,288371
15	MORACEA	2	0,13770026	19,789169
16	OCHNACEAE	1	0,00007854	3,3191769
17	RUBIACEAE	1	0,00125664	10,708403
18	SAPINDACEAE	1	0,1249917	2,3749599
19	SAPOTACEAE	2	0,0403138	13,380895

20	TILIACEAE	1	0,0290598	5,0023857
	total	**35**	**2,36679662**	**289,50049**

Table 13. Diametric distributions of the analysed forest

CATEGORIES DIAMETRIC	NO. TREES	VOLUME m3
I (1-9,9)	155	1,35786295
II (10-19,9)	44	4,45772023
III (20-29,9)	11	4,80575736
IV (30-39,9)	5	5,70464294
V (40 - 49,9)	2	4,2975957
VI (50 - 59,9)	1	3,45807222
TOTAL	**218**	**24,0816514**

Table 14. diameter categories by species

CATEGORIES DIAMETRIC	NO. SPECIES	NO. TREES	IVI
I (1-9,9)			
	Carob	7	22,5789011
	milkweed	1	2,54039561
	Anime	9	26,5699378
	Sawdust	1	2,63008242
	Caimito white	10	27,8328344
	caimito colorado	2	3,85152381
	charcoal burner	1	3,01872528
	cargadero colorado	2	6,75494507
	helmet	1	2,42081319
	knife	5	14,5854011
	Choiba	3	4,78397436
	Churimo	5	14,8672967
	coroba	1	2,54039561
	Guasca hoji wide	8	22,8266228
	Guasco colorado	3	8,65756777
	red guasimo	1	5,60967767
	Ant	16	37,5251887
	Soap	2	5,17805158
	Jagua	1	3,01872528
	black jigua	3	7,22899634
	white wool	2	7,99704764
	milkman	7	24,2131942

	lily	9	23,882881
	palo blanco	1	5,60967767
	swamp	4	13,2771429
	total	**105**	**300**
II (10-19,9)			
	Carob	3	29,8278782
	Anime	3	29,5442908
	Sawdust	4	37,7151522
	Caimito white	1	11,4669082
	carra	1	8,70193128
	knife	1	9,48179658
	Choiba	2	18,5559363
	Guasca hoji wide	1	8,70193128
	Guasco colorado	1	8,70193128
	red guasimo	1	12,0518072
	Ant	1	8,70193128
	Soap	2	21,3918101
	milkman	4	35,9256633
	lily	5	49,7492352
	swamp	1	9,48179658
	Total	**31**	**300**
III (20-29,9)			
	Carob	2	57,7914011
	milkweed	1	26,1894007
	Anime	3	82,7829314
	Sawdust	1	26,1894007
	Churimo	1	26,1894007
	milkman	1	26,1894007
	lily	2	54,6680649

	Total	**11**	**300**
IV (30-39,9)			
	Sawdust	2	116,600069
	Churimo	1	59,9104375
	Soap	1	59,9104375
	lily	1	63,5790561
	Total	**5**	**300**
V (40 - 49,9)			
	Sawdust	2	300
	Total	2	**300**
VI (50 - 59,9)			
	Choíba	1	300
	Total	1	**300**

FIGURES

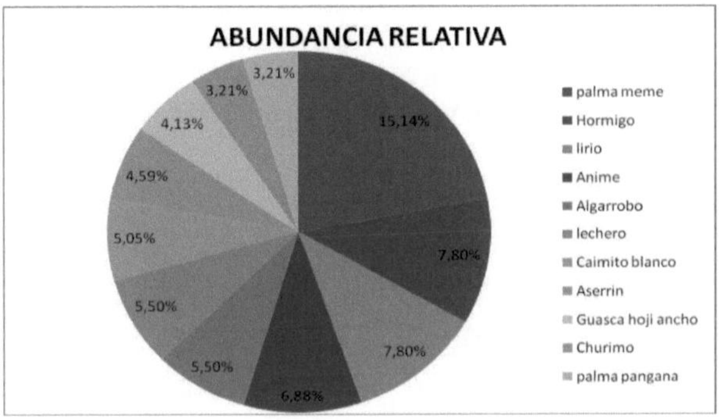
ABUNDANCIA RELATIVA
3,21%
3,21%
4,13%
15,14%
4,59%
5,05%
7,80%
5,50%
5,50%
7,80%
6,88%
palma meme
Hormigo
lirio
Anime
Algarrobo
lechero
Caimito blanco
Aserrin
Guasca hoji ancho
Churimo
palma pangana

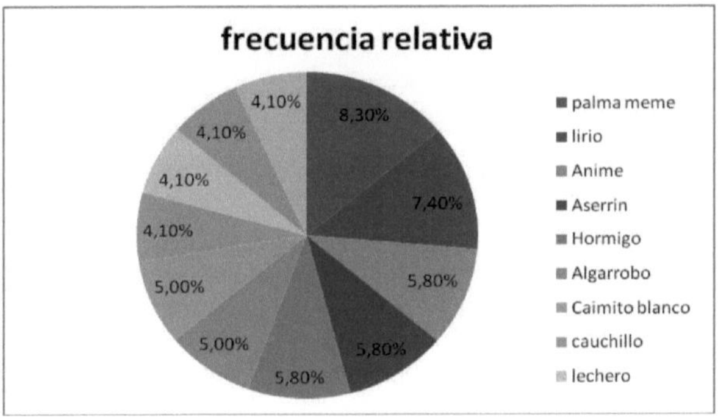
frecuencia relativa
4,10%
8,30%
4,10%
4,10%
7,40%
4,10%
5,00%
5,80%
5,00%
5,80%
5,80%
palma meme
lirio
Anime
Aserrin
Hormigo
Algarrobo
Caimito blanco
cauchillo
lechero

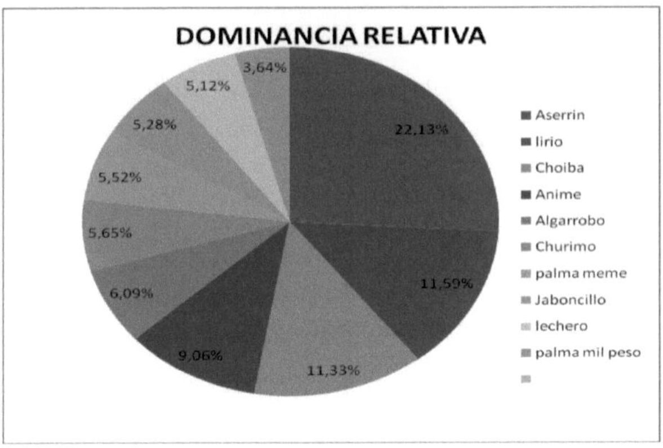
DOMINANCIA RELATIVA
3,64%
5,12%
5,28%
22,13%
5,52%
5,65%
6,09%
11,59%
9,06%
11,33%
Aserrin
lirio
Choiba
Anime
Algarrobo
Churimo
palma meme
Jaboncillo
lechero
palma mil peso

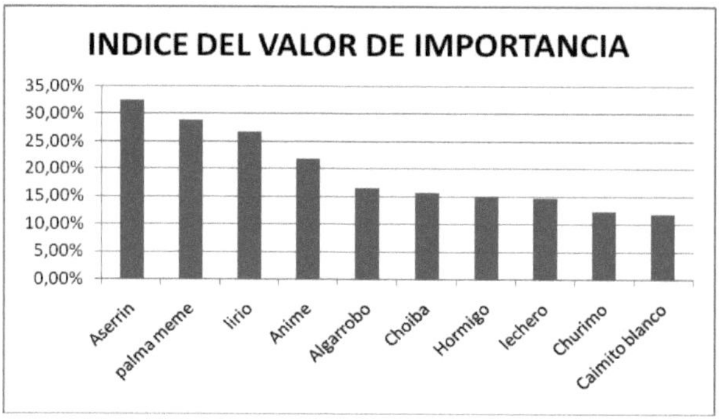
INDICE DEL VALOR DE IMPORTANCIA
35,00%
30,00%
25,00%
20,00%
15,00%
10,00%
5,00%
0,00%
Aserrin
palma meme
lirio
Anime
Algarrobo
Choiba
Hormigo
lechero
Churimo
Caimito blanco

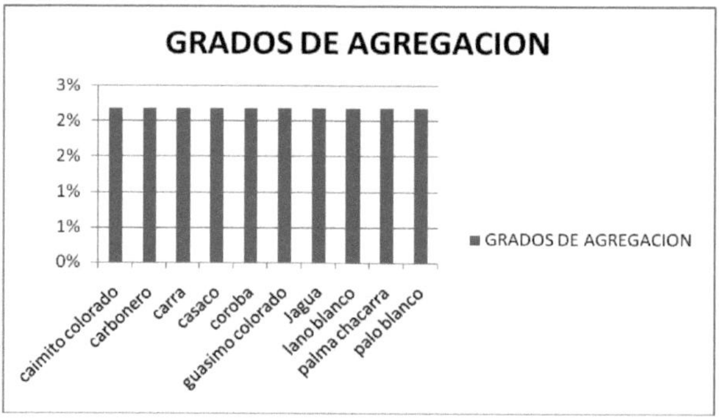
GRADOS DE AGREGACION
3%
2%
2%
1%
1%
0%
caimito colorado
carbonero
carra
casaco
coroba
guasimo colorado
Jagua
lano blanco
palma chacarra
palo blanco
GRADOS DE AGREGACION

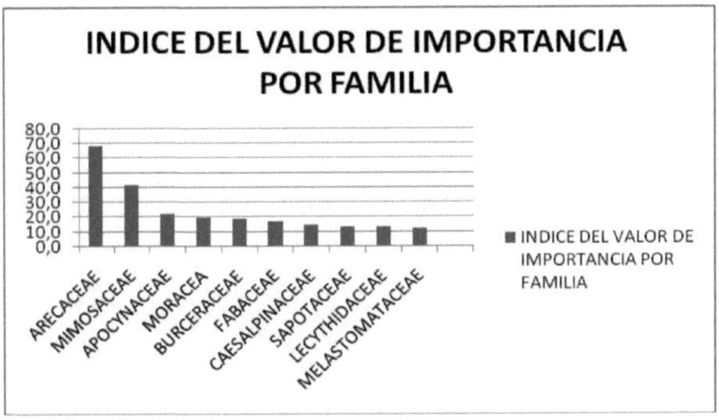
INDICE DEL VALOR DE IMPORTANCIA POR FAMILIA
80,0
70,0
60,0
50,0
40,0
30,0
20,0
10,0
0,0
ARECACEAE
MIMOSACEAE
APOCYNACEAE
MORACEA
BURCERACEAE
FABACEAE
CAESALPINACEAE
SAPOTACEAE
LECYTHIDACEAE
MELASTOMATACEAE
INDICE DEL VALOR DE IMPORTANCIA POR FAMILIA

OGAWA INDEX

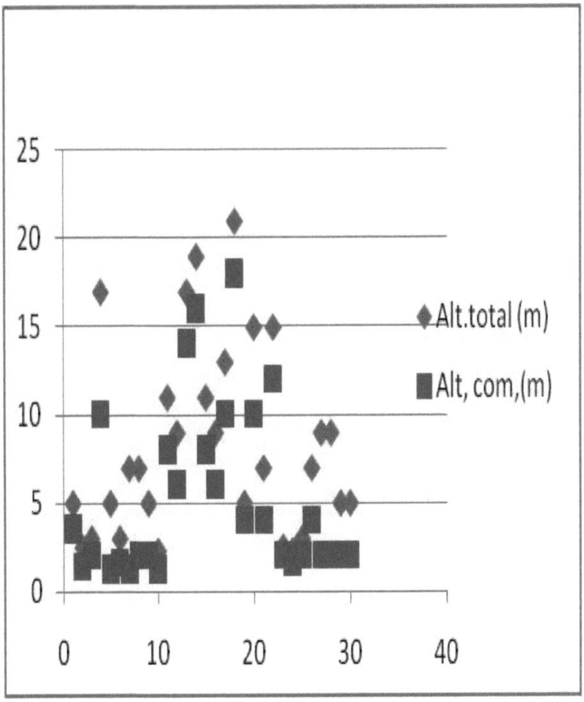
25
20
15
10
5
0
0
10
20
30
40
Alt.total (m)
Alt, com,(m)

	Total height (m)	Commercial height (m)
pangana palm	5	3,5
Anime	2,5	1,4
Choiba	3	2
Carob	17	10
Coroba	5	1,2
palm meme	3	1,6
white alligator	7	1,2
Ant	7	2
Sawdust	5	2
guasca colorado	2,3	1,2
white alligator	11	8
palm meme	9	6
palm meme	17	14
Anime	19	16
Anime	11	8
Anime	9	6
Anime	13	10
Choiba	21	18
caimito colorado	5	4
Lily	15	10
Ant	7	4
Swamp	15	12
caimito colorado	2,5	2
Carob	2,3	1,6
Choiba	3	2
Choiba	7	4
Choiba	9	2
Sawdust	9	2
Ant	5	2
Lily	5	2

MIX
Papier aus verantwortungsvollen Quellen
Paper from responsible sources
FSC® C105338

Printed by Books on Demand GmbH, Norderstedt / Germany